NEOCLASSICAL

ARCHITECTURE & DESIGN LIBRARY

NEOCLASSICAL

Young Mi Kim

FRIEDMAN/FAIRFAX
PUBLISHERS

A FRIEDMAN/FAIRFAX BOOK

©1998 by Michael Friedman Publishing Group, Inc.

All rights reserved. No part of this publication may be reproduced, stored in a retrieval system, or transmitted, in any form or by any means, electronic, mechanical, photocopying, recording, or otherwise, without prior written permission from the publisher.

Library of Congress Cataloging-in-Publication Data available upon request

ISBN 1-56799-434-2

Editor: Reka Simonsen
Art Director: Jeff Batzli
Layout Design: Jennifer Markson
Photography Editor: Wendy Missan
Production Manager: Camille Lee

Color separations by Colourscan Overseas Pte Ltd.
Printed in Hong Kong by Midas Printing Limited

10 9 8 7 6 5 4 3 2 1

For bulk purchases and special sales, please contact:
Friedman/Fairfax Publishers
Attention: Sales Department
15 West 26th Street
New York, New York 10010
212/685-6610   FAX 212/685-1307

Visit our website:
http://www.metrobooks.com

For Scott, whose love and patience bring out my best, and for my loving family: Mom, Dad, Young Bae, Heung Bae, Eumene, and little Spencer.

Contents

INTRODUCTION
8

CHAPTER ONE
ARCHITECTURE
18

CHAPTER TWO
INTERIORS
40

CHAPTER THREE
FURNITURE
60

CHAPTER FOUR
ACCENTS
74

INDEX
96

INTRODUCTION

Like fashion's little black dress, Neoclassicism is ever in style. Rooted in immutable Greek and Roman antiquity, Neoclassical style is often described by designers and decorators as an enduring, unmistakable aesthetic—a classic. Yet for all its ancient appeal, the popularity of Neoclassical today is due more to its ever-changing nature: designers from every corner of the Western world have reinvented the style again and again to suit each new period, proving that Neoclassical style is not just timeless, but also universal.

Ironically, the style that is considered such a classic today was on the wane in the early eighteenth century, but when the ancient city of Pompeii was unearthed in 1748, interest in all things classical came roaring back. One character who figured prominently in reviving—and remaking—classical design was Napoleon Bonaparte. In fact, scholars have described the beginning of Neoclassical style as a celebration of Napoleonic ideals of imperialism.

During his reign in the early 1800s, Napoleon mandated a design that was antithetical to the excessively decorative Rococo style, which was favored by the Imperial Courts of Europe. Borrowing heavily from ancient Greek civilization, as well as from ancient Roman, Egyptian, and Etruscan art, craftsmen of the period created a style whose best motifs—columns, capitals, pedestals, mythological creatures, and stylized plants and animals—were not new but were newly idealized.

Neoclassical style offered a more refined, if less exuberant, philosophy to decorating. If Rococo was all flowery opulence, then Neoclassical was orderly, dignified, and linear. Indeed, the aesthetic of Neoclassical style offered perhaps the first modern example of the "less is more" school of decorating, with its fondness for simple, geometric forms, spare use of Greek and Roman architectural elements—particularly columns and pedestals—and restraint in the use of color and decorative objects.

The first interior design incarnation of the Neoclassical aesthetic, French Empire style, was certainly more grandiose than the style it

OPPOSITE: *This conservatory has a Neoclassical focal point: behind the marble statuary stands the facade of a Greek temple. Two fluted Doric columns support an elaborate pediment, which contains dentil molding within the tympanum (the triangular section of the pediment).*

superseded (Rococo), but it was also more refined. It had an orderly and dignified approach to furnishings that was lacking in the earlier style. And it was new: never before had designers conjured a style balancing rectilinear designs with ornate symbols and motifs. The result was strikingly modern for its time.

Although most closely associated with Napoleon, the style actually owes its birth to two artisans whom Napoleon commissioned to decorate Malmaison, the palace of his wife, Josephine: Charles Percier and Pierre-François-Léonard Fontaine. Percier and Fontaine set about creating an interior design that was emblematic of an emperor, using Napoleonic symbols such as bees, giant letter Ns in laurel wreaths, and eagles, as well as forms borrowed from ancient Rome and Greece. They also introduced another element that became the linchpin for their new Empire style. Following Napoleon's campaign in Egypt from 1798 to 1799, Percier and Fontaine began incorporating Egyptian symbols into their design. Lion's heads, lotus and papyrus flowers, palmettes, reeds, and sphinxes added exotic touches to the otherwise Western look.

While the influence of Empire style was growing—and changing—with each of Napoleon's conquests, another decorative style similar to Empire was taking hold across the Channel in England. Historically, the Regency refers to the period from 1811, when George, Prince of Wales, became regent of England, to 1820, when he became King George IV. As a decorative style, however, the period is actually

RIGHT: *This elegant dining room is the quintessence of Neoclassical style. The ceiling bears a symmetrical plasterwork design that incorporates a classical garland motif, as does the mantelpiece. The door's lintel supports a broken pediment, and the doorway itself is flanked with pilasters.*

much broader, describing the interior design and furnishings predominant in England from the 1790s to the early part of Queen Victoria's reign in the 1840s.

Like French Empire, English Regency was a reaction against the delicately ornate period that preceded it, echoing vestiges of Baroque and Rococo. And the designs were obviously influenced by the French Empire style, minus the pompous Napoleonic ornamentation. In fact, the most distinguishing characteristic of the Regency style was that furnishings tended to be more ample and sturdy-looking—and decorated with fewer opulent details—than those of Empire.

With Regency style, the English toned down Empire style a notch. A similar trend was under way in Germany, creating yet another offshoot of Neoclassical style. Beginning in 1810 and lasting into the 1860s, Biedermeier style reigned in Germany, Austria, and to a lesser extent Scandinavia. Thought to embody the bourgeois way of life in those countries, "Biedermeier" refers mainly to the furniture of the period, which possessed an unpretentious elegance in its solid, comfortable-looking designs. Biedermeier, in fact, was one of the truly popular aesthetics that sprang up in post-Napoleonic Europe, becoming the style of choice for the expanding middle class. The clean, simple lines and use of light-colored woods gave Biedermeier furniture pieces distinction—without the bravado of Empire style.

In America, the face of Regency style changed as rapidly as the new democracy spread. With furnishings and architecture drawn largely from English designs, Federal style was ushered in while the aesthetic of the Colonial era was on its way out. The Federal period, which lasted from the establishment of the federal government in 1789 until 1830, was marked by its reliance on English sources.

Architect Robert Adam, whose highly ornate designs based on Greek and Roman antiquity had swept through England prior to the advent of Regency style, naturally became the preeminent inspiration for architecture in America during this period. English craftsmen such

OPPOSITE: *This room pays homage to both Eastern and Western design traditions. The porcelain jars and magnificent cabinet recall the Victorian passion for Chinoiserie, while the X-bench, framed prints of Greek architecture, and beautiful marble column are purely Neoclassical.*

ABOVE: *With a fluted shaft that is distressed to create an "ancient" finish, this column is a show-stopper in a room that already has plenty of classical flair. From the stone floor to the wide linear moldings and cream-colored walls, the room has the open, airy feeling of a courtyard. Plasterwork ornaments and statuary add to the effect.*

INTRODUCTION

as George Hepplewhite and Thomas Sheraton—both of whom called themselves cabinetmakers but were best known for their design books—became the definitive furniture influences behind Federal style, alternately described as American Empire.

Regardless of its incarnation, though, Neoclassical style has a unique signature, combining classically inspired forms with detailing that is often ornate. It is a look that architects, designers, and homeowners have found intriguing time and again; Neoclassical, it would seem, offers something for everyone.

ABOVE: *This Ionic column, with its scrolled capital and fluted shaft, gives this room unmistakable Neoclassic flair. Light walls set off only by white trim—including an unusual built-in ledge acting as the room's ceiling molding—enhance the simplicity and refinement of the room's architectural style.*

ABOVE: *Providing stark relief against a boldly painted wall, a pair of Ionic columns are the ideal foil for this room: not only do they, along with the entablature they support, create a sense of architectural importance, but they are also remarkably at ease in a modern setting.*

OPPOSITE: *Whether they're made of marble, granite, wood, iron, or concrete, columns are the most persistent of architectural elements, representing Greek and Roman civilizations in a single form. According to the rules created by ancient architects, columns rarely stand alone, except when used for certain memorial purposes. Here, a typical arrangement of twin Tuscan columns forms the support for a series of arches.*

INTRODUCTION

INTRODUCTION

ABOVE: *Flanking a room filled with English Regency pieces, the two entrances display Neoclassical motifs such as garlands, ribbons, and swags in the friezes of the wide entablatures. Slender Tuscan columns give further definition to these doorless entrances.*

OPPOSITE: *Demonstrating the prestige of imperial France, French Empire furnishings and accessories were often heavily gilded. In this dining room, the gilt frame and gilding on the marble-topped console enhance the luster of the mahogany table. Period accents such as the bronze sphinx were also used frequently during Napoleon's reign.*

CHAPTER ONE
ARCHITECTURE

Today, designers and homeowners who choose to decorate in Neoclassical style have a rich environment in which to create. Many of the best-known public and private buildings in eighteenth- and nineteenth-century Europe and America were designed with classical ideals in mind.

One of the most commonly employed of these ideals was the column. Although Egypt was probably the first civilization to incorporate columns into building architecture, the ancient Greeks established an "Order," or a design aesthetic, that laid out basic principles for usage. Created by the Greeks between the seventh and fourth centuries B.C., the basic Orders of classical architecture—Doric, Ionic, and Corinthian— have been used by architects throughout the ages. (The Romans added the Composite and Tuscan orders later.)

Beginning with the Doric Order, the plainest of the column shapes, the Greeks soon defined the more embellished and organic shapes of their other two Orders. The Ionic Order has a scrolled capital, while the Corinthian Order has a tall bell of acanthus leaves terminating in diagonally projecting small scrolls, or volutes.

Originally created as functional elements, columns today are more symbolic, representing in a single concept the power of Western civilization. Churches, temples, banks, government buildings, and houses in Europe and America frequently have columns, pedestals, and arches as part of their architecture.

Houses designed in the Neoclassical fashion often featured columns as part of their facade, as well as incorporating them into interior spaces. In these houses, decorative molding was requisite to line the top of a wall or surround a niche. Engaged columns, which project from a solid wall in a semicircular shape, and pilasters, which are flat and project less than engaged columns, are other quintessential ele-

OPPOSITE: *Georgian style reflected Renaissance ideals that were revived in England during the early 1700s. Characteristics of the style include rigid symmetry, hipped roofs, and sash windows, all of which are exhibited in this grand house. Another Neoclassical feature is the two-story portico supported by massive Tuscan columns.*

ments that architects often relied on to break up an otherwise uninteresting expanse of wall or to define an opening.

Leading architects of the eighteenth and nineteenth centuries adapted these and other classic forms to suit new eras, among them Robert Adam, a British architect whose interior designs have come to epitomize Neoclassical ideals in architecture and interior design.

Adam set the stage for Regency style in England with his lasting designs for houses and other buildings, interiors, and furniture. With his adaptation of Etruscan style, which incorporated such classical ornamental motifs as palmettes and lotus flowers, Adam revived interest in the ancient cultures of the West. But even as Adam's style swept throughout England in the late eighteenth century, his designs became increasingly exaggerated and excessively decorated—the opposite of what Regency represented. Adam's embellished style fell into disfavor just as the simpler Regency style was coming into its own.

Adam's designs were inspired in large part by the works of Andrea Palladio, a prominent architect of the sixteenth century. These Palladian designs, as interpreted by Adam, generally called for a massive scale, with the use of elements such as pediments, cornices, lion masks and paws, acanthus leaves, and swags, all of which were derived from architectural ornament or antique sculpture. In fact, most architects of western Europe and America during the eighteenth century became obsessed with re-creating the look of a Greek temple. The legacy is apparent in both public and private buildings that are still standing today.

Many of those buildings were erected in America, from New York to Georgia, in the 1830s and 1840s during a stylistic period known as Greek Revival. Perhaps because America was a newly minted democracy—and enthralled with the idea of being the spiritual successor to ancient Greece—the fervor for all things classical came about at this time (even towns throughout the country were named after Greek capitals such as Athens, Sparta, and Ithaca). Led by Benjamin Henry Latrobe, an architect and engineer who designed and built the Bank of Pennsylvania—the first American building to incorporate a classical Greek Order—the Ionic-Greek Revival struck a chord.

The look of a Greek Revival house is unmistakable. Most often featuring columns and pilasters, buildings in this style usually have bold, simple moldings decorating both exterior and interior; pedimented gables; heavy cornices with unadorned friezes; and horizontal

LEFT: *Highly popular in the late eighteenth century, the Greek key motif was often found in railings, balustrades, entablatures, and moldings. Its clean, linear design is immediately recognizable from antiquity, yet the same spare quality works equally well in contemporary settings.*

transoms above entrances. Also, many public buildings possess impressive portico entrances that have two or more massive column supports.

Of course, Greek Revival isn't the only classically inspired style of architecture; it's simply the most obvious. Under the Neoclassical umbrella, architectural styles such as Georgian, Federal, Jeffersonian, Italianate, Second Empire, Beaux Arts, and Classical Revival are also strongly rooted in Greek and Roman aesthetics.

With the architectural emphasis in much of the Western world focused on classical models, it's no wonder that Neoclassical style came to the forefront of interior design. Very few styles, past or contemporary, so aptly fit the imposing structures of classical buildings and homes.

BELOW: *Delicate plasterwork details on ceilings and walls were an important part of the interior architecture of homes built during the height of Neoclassical style in England. This ceiling detail shows the influence of architect Robert Adam, who favored elaborate but symmetrical detailing in moldings: a florette is at the center of a sunburst design, which is edged with garlands and oval medallions.*

ABOVE: *Nothing says Neoclassical quite like a Palladian-style window, a tripartite design that features a large arched central window and flanking rectangular side windows. Here, the design is embellished with a plain broken pediment atop four pilasters (shallow piers attached to walls), and small rectangular window panes highlight the building's geometry.*

ARCHITECTURE

ABOVE: *During the 1830s and 1840s, the Greek Revival style flourished in the United States, leaving behind a legacy of houses that look like miniature Greek temples. The most easily identifiable features of these houses are the columns and pilasters, although not every Greek Revival house has them. Other hallmarks of the style are represented in this house: bold, simple moldings both inside and out, pedimented gables, heavy cornices with unadorned friezes, horizontal transom windows above entrances, a portico, and a bright white facade.*

OPPOSITE: *This house shows its Neoclassical pedigree with a number of architectural elements. The smooth Tuscan columns are topped by saucer-shaped capitals and support a bold dentil-molding design in the cornice. The design is repeated in the tympanum (the triangular area) inside the pediment.*

ARCHITECTURE

OPPOSITE: *During the Regency period in England, wealthy homeowners used Neoclassical design to create "the country in the city." This gazebo achieves the ideal beautifully, acting as the focal point for a lush garden setting. Formed with five modified Corinthian columns, the gazebo is capped with a circular entablature that features a frieze of carved swags.*

ABOVE: *Used outdoors, Neoclassical elements inevitably capture the beauty and spirit of an Italian garden. Here, a Corinthian column—characterized by a slender fluted shaft and a capital decorated with stylized acanthus leaves that end in volutes—stands in contrast to the simple pergola it helps to support. Oversized stone planters shaped like urns add to the garden's classical look.*

ARCHITECTURE

ABOVE: *An otherwise ordinary veranda is transformed by the ornate Corinthian capitals atop these fluted columns. An elaborate balustrade along the bottom and arches at the top separate the columns, creating a grand arcade effect. Simple wooden shutter doors add color and contrast to the entrance.*

ABOVE: *Columns can add structure and Neoclassical beauty to any garden setting. Here, unadorned Tuscan columns form the gateway to a charming patio, where guests can dine alfresco. The brick floor and stone walls surrounding the property accentuate the feeling of a private retreat.*

ARCHITECTURE

ABOVE: *A stark white pergola creates the framework for this Neoclassical garden design. An arcade is created by the rows of simple Tuscan columns that line the pergola; flagstone tiles delineate the space further. Terra-cotta planters and climbing vines add color and life to the scene.*

ARCHITECTURE

ABOVE: *The cool, serene look of marble works beautifully in this room to capture the spirit of Neoclassical ideals in a contemporary framework. Two truncated marble obelisks (an ancient Egyptian design that became a symbol of Napoleon's empire), each supporting a classically shaped Greek urn, turn this entryway into a dramatic focal point worthy of an ancient temple.*

ARCHITECTURE

ABOVE: *Set to a modern tempo, Neoclassical style assumes even greater importance in a room. Here, plain Tuscan columns flank a pair of tall French doors, which came into use in the late eighteenth century. Adding to the architectural dimension of the space are the subtle wainscoting and molding and the structured walls that set off the French doors.*

RIGHT: *This minimalist foyer depicts one of the best features of Neoclassical style: contrast among elements. Cool white marble blocks and the matching column provide the quintessential classical backdrop, while accents—such as the bust and pedestal table—are kept to a minimum.*

ABOVE: *A simple, arched doorway shows a glimpse of antiquity in the foyer beyond it: a marble-topped Empire-inspired console table sets the mood for the entry area. The resplendent gilt mirror hanging above the table highlights the golden accents found on surrounding items, such as the sconces, clock, and candlesticks.*

ABOVE: *Sitting atop pedestals, these classic Doric columns endow this home's interior architecture with Neoclassical personality. Enhancing the setting are furnishings typical of the period: an upholstered X-bench sits in the hall under a drawing of a Greek urn, and the dining room features American Empire furnishings.*

ARCHITECTURE

ABOVE: *Marking the entry to the dining room, these two Tuscan columns support an arch with a plain keystone at the top. The simple lines and plain colors of the columns and arch set the mood for the room, which is filled with an eclectic, pared-down mix of furnishings and accents.*

ARCHITECTURE

ABOVE: *Geometric paving was widely adopted during the early Georgian period (early- to mid-1700s) in England, when Neoclassical ideals were gaining popularity. Here, a typical floor from the era shows a pattern of equal black and white squares, set diagonally. Plain Tuscan columns supporting an entablature with a modified dentil molding also attest to this home's classical lineage.*

OPPOSITE: *Soft golden walls offer an ideal backdrop for contrasting classical ornamentation, such as the framed Piranesi-like etchings, the "torch" wall sconces, and the whimsical lettered border that runs along the top of the wall. An area rug edged with a Greek key border further distinguishes the room.*

ARCHITECTURE

ABOVE: *Architects have used columns in combination with arches for ages, creating a kind of rhythmic vertical organization within a house. Here, two arches, each supported by marble Tuscan columns, provide a striking transition from the dining area into a living space. To draw the eye up, ceilings and moldings are painted light colors.*

OPPOSITE: *In this early Georgian dining room, the walls are the primary Neoclassical attraction. Painted a sunny yellow with contrasting white trim, the walls are divided into three sections, typical of Georgian style: frieze (a panel below the upper molding or cornice of a wall), field (the upper part of a wall, below the cornice), and dado (the lower wall surface, from the chair rail down to the baseboard).*

ARCHITECTURE

ARCHITECTURE

OPPOSITE: *Neoclassical style is eminently suitable for a modern setting, as this interior demonstrates. Large white cinder blocks set an industrial mood for the dining area, where a plain table sits flanked by two wide, X-back chairs—pared down from the original Empire design.*

ABOVE: *Although the vaulted ceilings are more evocative of Romanesque style, this kitchen nonetheless bears the unmistakable stamp of Neoclassical style. Rich, light-colored maple, shown in the cabinetry here, was a favorite wood of Biedermeier cabinetmakers during the early- to mid-nineteenth century. The floors, too, exhibit classical tendencies: the design, a modified version of the original* carreaux d'octagones, *was popular throughout the eighteenth century.*

ARCHITECTURE

ABOVE: *This homeowner created a miniature Roman bath by using simple classical elements. Marble columns in the Tuscan style flank the tub enclosure, defining the bathing area as the room's focal point. Whitewashed brick walls, marble statuary, and a neutral palette finish the room in typically tranquil contemporary Neoclassical tones.*

OPPOSITE: *Ribbed pilasters support an elliptical arch, creating a striking tub alcove in this bathroom. Pilasters lend an implied structure and rhythm to a space, and architects use them to define an opening in a wall or to break up and delineate large expanses of wall surfaces.*

CHAPTER TWO
INTERIORS

Although based on formal rules of Greek and Roman architecture, Neoclassical isn't just for purists. Today, decorators, homeowners, and aficionados of antiquity have incorporated elements of Neoclassical into their interiors, creating new ways to approach a classical style.

Napoleon, who crowned himself emperor in 1804, commissioned two designers to decorate Josephine's palace, Malmaison, in the style of Greek and Roman antiquity—but personalized with Napoleonic overtones. French architects Charles Percier and Pierre-François-Léonard Fontaine, already great admirers of ancient cultures, accomplished this goal by eschewing the Rococo style that was so popular at the time. Instead, they developed their own version of classicism—French Empire style, a tribute to both the classical past and to Napoleon's recent victories in Italy and Eygpt—that has endured to this day.

Although the style is recognized for having a foundation resting on Grecian ideals, the motifs used in Empire style came from many sources. Egyptian decorations such as the sphinx and the winged lion, Greek caryatids, and many symbols such as the swan (Josephine's emblem was the black swan), the imperial eagle, stars, palmettes, lotus flowers, medallions, trophies of arms, and Napoleonic bees are heavy with mythological meaning, which is undoubtedly the reason Napoleon's designers chose them.

Combining Napoleon's imperial symbols with motifs from Greek, Roman, and Egyptian civilizations—three of the world's most powerful empires—this composite style worked beautifully, and it's no wonder. Ancient Greeks actually borrowed heavily from Egypt, and ancient Romans had mined Greek culture and art. Napoleon's co-opting of symbols from antiquity actually renewed and reinvigorated a classic look.

But more change was on the way. One English architect was so taken with Empire style that he simply appropriated it, bringing it to

OPPOSITE: *This ebonized secretary is the centerpiece for a Neoclassical vignette. Elements that recall the refinement of Greek and Roman art forms are placed against an ocher-colored wall, which features accents evocative of the Empire style, including swags, tassels, and a gilt frame.*

England in a startlingly high-profile job: architect Henry Holland filled the Prince of Wales' Carlton House almost entirely with French Empire furnishings. In the new environment, the furniture took on a different air—and a new name: Regency.

The newly assimilated style was actually more reserved than French Empire, having more heft and less embellishment. For the English, Regency was right for the time, expressing an opulent sensibility not marred by the garish ornamentation of Baroque or Rococo, the preceding styles. And the Regency style admirably suited the period's penchant for classical architecture, brought about largely by architect Robert Adam.

Another thing that set the style apart from Empire was that interiors decorated in the Regency style were much more arranged. Whereas a French drawing room's furniture was almost inevitably placed against the walls, to be moved only for social occasions, a Regency interior often had one or more groupings of furniture that emphasized a natural focal point of the room, such as the fireplace. In this way, the English were the first to create "conversation pits," a term that designers use today to define close groups of furniture—with sofas or chairs facing each other—that facilitate social interaction.

In design, England had its own version of Percier and Fontaine to trumpet the more spare ideals of this new Regency style. Thomas Hope was a wealthy banker (and close friend of Percier and Fontaine) who liked to travel to locations such as Greece, Egypt, Syria, and Sicily, where he could feed his love of antiquity, but he was not a designer or decorator. Yet from his travels, Hope had cultivated a remarkable sense of interior design, which he drew upon to write a book that became one of the cornerstones of Regency style, *Household Furniture and Interior Decoration*. A year later, another influential treatise, *Collection of Designs for Household Furniture and Interior Decoration*, was published about the Regency style, this time by noted cabinetmaker George Smith.

Both books were filled with Neoclassical visions of the home, set against the backdrop of early-nineteenth-century architecture. But even in a modern setting, Neoclassical doesn't disappoint. With fantastic imagery taken straight out of Greek and Roman mythology and mixed with the majestic refinement of columns and bas-relief carvings, Neoclassical aims simply to inspire.

With this versatility in mind, contemporary architects and designers have decorated everything from ultramodern lofts to suburban ranch houses in the Neoclassical style. In a time when many homes are built with little attention paid to architectural features, Neoclassical style assumes even greater relevance, adding depth, shape, and order—as well as quite a bit of character—to a home's interior.

OPPOSITE: *Set against a colored wall, the plaster molding of this door surround stands out as an unmistakable example of Empire-style craftsmanship. The bracketed console contains an elaborate frieze of palmettes and vines while the top of the doorhead culminates in cresting and a central design of scrolling vines and shells.*

ABOVE: *The decorative painting technique known as trompe l'oeil is used to great effect in this vestibule. Although it's difficult to tell with some items, most of the classical elements here are pure artifice: a real black-and-white tile floor continues in a mural that depicts a balcony opening onto a pastoral landscape, while the "marble" walls, niche, and molding are painted on.*

OPPOSITE: *Simpler patterns in flooring came into vogue during the late Georgian period in England, with marble floors such as this one often used in vestibules and baths. The black squares against white marble tiles form a precise geometric pattern that is offset in typical Neoclassical fashion by the ornate stone capital (used here as a table base). The heavy gilded mirror boasts another Empire symbol: the eagle.*

OPPOSITE: *Neoclassical design is both functional and decorative. In this apartment, a wall shaped to resemble a Greek temple facade acts as a room divider, separating the dining area from the kitchen. The wall, which features a pediment and entablature—although no columns—creates a dramatic focal point in a room that has little in the way of architectural features.*

ABOVE: *The trend toward moving furnishings away from the wall to be centered around a hearth began during the Regency period. Here, a modern adaptation still works. The columns act as a screen separating the hall from the living room; the spare detailing of the Tuscan columns echoes the architecture of the room.*

ABOVE: *A folding table and chairs provide a whimsical lift to this corner. The tongue-in-cheek 3-D cutout depicting Piranesi-style sketches of a Roman bath acts as a base for the table; the black and white scheme of the geometric tile floor makes the Neoclassical flavor even more pronounced.*

ABOVE: *There's no telling where Neoclassical will turn up and make itself at home. In this pristine futuristic setting, the style easily adjusts to the tempo of the room. From the modern interpretation of an Empire-period desk and the lyre-backed armchair—an archetypal Neoclassical piece—to the modified Ionic column, the room has a classical personality all its own.*

ABOVE: *The decorative serenity of Neoclassical style is apparent in this modern arrangement, which uses carefully chosen appointments and a monochromatic scheme to work its magic. A gilded X-bench in the foreground and the scroll-back armchair openly embrace classical aesthetics, while a classical bust sits amid examples of a newer art form—photography.*

INTERIORS

OPPOSITE: *With its mix-and-match furnishings, ranging from Shaker chairs to an antique chest, this living room is certainly eclectic. Yet Neoclassical style is unmistakably the glue that holds it all together. Doric columns, elegant moldings, and Empire-period furnishings all attest to the room's classical leanings.*

ABOVE: *With the lure of a formal garden just outside its French doors, this room offers the refined, relaxing atmosphere for which Neoclassical interiors are famous. Oversized, tufted scroll-back chairs in rich silk jacquard, a stone bust, and the ornate urns placed on each side of the French doors lend Regency-style grace to the setting.*

ABOVE: *A combination of classical architecture and period furnishings creates this room's Neoclassical look. The rich color of the wood of the Regency-style sofa contrasts handsomely with the sumptuous silks of the upholstery and decorative roll pillow. These types of sofas were popular in drawing rooms during the period.*

OPPOSITE: *In this formal dining room, Neoclassical style fits in effortlessly: the crown molding and the rich ocher walls lend a regal, Empire-period look to the space. Against this backdrop, a delicate green reminiscent of patina graces wainscoting and the fireplace, which features Greek motifs in relief.*

ABOVE: *An arched pediment with an elaborate frieze and two Greek urns create a dazzling Neoclassical ensemble in this bedroom. Sitting atop an enormous picture window that is flanked by two narrower casement windows, the classical architectural elements are decoration enough for the wall. The bed, however, is outfitted in typical French Empire fashion, with layers of soft fabric hung tent-style above the headboard.*

INTERIORS

ABOVE: *The cool, elegant look of a black and white marble floor provides an appropriately Neoclassical foundation for this bathroom. At once luxurious and understated, the spare decor of the room underscores the beauty of simple effects, from the sketch of an urn to the rich wood siding of the tub. Sunny yellow walls and curtains set the scene aglow.*

INTERIORS

ABOVE: *With its miniature-tile floor and tub surround, this is one bathroom the ancient Greeks and Romans would no doubt have admired, while the Empress Josephine would have recognized the elegant oval of the tub. Of course, elements such as the plain columns flanking the tub entrance and the intricate mosaic tile design in the center of the floor enhance the contemporary Neoclassical feel.*

RIGHT: *This homeowner used the clean, balanced look of certain Neoclassical elements to create a fanciful contemporary setting. Two freestanding Ionic columns flank the foot of the bed, where an aluminum chair and bistro table sit dreamlike, reflecting the "sky"—the ceiling and walls invent a celestial scene indoors.*

ABOVE: *In this marble bath, two gilded X-benches are the only furnishings necessary to convey the Neoclassical aesthetic. The form, which comes from the ancient Greeks, was revived during the Empire and Regency periods and was used most frequently in vestibules, drawing rooms, bedrooms, and bathrooms. This adaptation of the ancient form is purely modern.*

OPPOSITE: *Building on its already classical architecture, this bath's Neoclassical personality shows in the trompe l'oeil frieze on the tub and just below the ceiling molding. Heavy draperies are drawn back to enhance the mural in the background; Napoleonic devices such as swags, birds, and arrows, along with the geometric tile floor design, are richly evocative of French Empire style.*

ABOVE: *Trompe l'oeil effects work wonders in this bathroom, where an Italian landscape seen through archways has been painted above a (real) marble ledge. Setting a relaxing mood, the painting achieves its illusory goal by using tones consistent with that of the marble. Trompe l'oeil was a favorite Neoclassical technique among artisans of the eighteenth century, though it is just as appropriate in a modern setting.*

CHAPTER THREE
FURNITURE

Greek and Roman antiquity may provide the refined, dignified framework for Neoclassical style, but the eccentricities that represent the "neo" in the term furnish the personality. Sphinxes, carved animal figures, and draped female figures—bronzed, gilded, or made of marble—are signature Neoclassical elements used in furniture. Tempered by the orderly aesthetic of Western classicism, these odd, over-the-top elements are refreshingly daring to modern eyes, often adding wittily irreverent accents.

It must have seemed that way to Charles Percier and Pierre-François-Léonard Fontaine, who, as the architects of French Empire style, had a taste for the exotic both in their choice of motifs and in the materials they used. Employing France's best cabinetmakers, notably Georges Jacob and later his sons, to build the furniture, Percier and Fontaine commissioned pieces that were extravagant in their use of material, yet elegant in their clarity of line. Usually adapted from classical forms, Empire pieces were often simple, even geometric. But with lavish gilded details such as winged lions, swans, and sphinxes drawing much of the attention, the pieces needed the classical grounding.

Although Percier and Fontaine would have preferred exotic woods such as mahogany or rosewood for the furniture, the Napoleonic Wars made it difficult for craftsmen to obtain these and other treasured materials: instead, they relied on native woods such as burl maple, beech, ash, and walnut. Craftsmen of the period also relied on unusual cabinetmaking techniques to add embellishment. Along with contrasting wood inlays, furniture makers incorporated metals such as silver, gold, pewter, and steel and luxurious materials such as mother of pearl and marble into designs whenever possible.

Meanwhile, across the English Channel, Regency furniture was being designed with the same eye for linear integrity and exotic ornamentation. In fact, Empire furnishings popular at the time underwent only subtle adaptations to achieve the Regency look. George Hepplewhite's posthumously published book, *The Cabinetmaker and Upholsterer's Guide*, shows the classically inspired, elegant look upon which

OPPOSITE: *This elegant sofa, with its unusually shaped back rail, reveals the lasting impact of the Neoclassical movement. Based on an English late-Georgian mahogany sofa design, the lower scrolled end was a characteristic trait of these types of sofas. The exaggerated cabriole legs became fashionable in the 1930s in America, while the use of silks—in this case a self-colored silk woven with alternating matte and shiny stripes—has always been in vogue for these sofas.*

ABOVE: *Sphinxes were a frequently used device in French Empire furnishings after Napoleon returned from his successful campaign in Egypt. Today, sphinxes are one of the most lasting motifs of Neoclassical design. Sitting atop this bookcase, a pair of facing sphinxes are the crowning element in this classical vignette, which also features an urn-shaped lamp, an Empire-period console with brass fittings, and a gilded mirror.*

Regency would be based. Essentially, it was a sourcebook of French Empire designs, stripped down slightly.

Today, the Regency style is known best by the talented craftspeople, particularly cabinetmakers, who designed the furniture—Sheraton and Chippendale being the most famous. Unlike the plight suffered by cabinetmakers in France, craftsmen in England were able to obtain the most exotic woods possible for their designs, including mahogany, rosewood, zebrawood, amboyna, and maple, which was used mainly for veneers.

Homeowners today who prize Regency design but can't afford the small fortune charged for original pieces from the period have benefited from the ongoing popularity of this Neoclassical style. Furniture makers continue to produce pieces that, because of their classical pedigree, sit well in any setting.

The variations of Neoclassical style didn't stop with Regency. In the latter part of the Federal period, a style sometimes referred to as American Empire came about just as Napoleon's empire in Europe was crumbling. Unlike its French cousin, American Empire was not infatuated with ornamentation. Although early pieces display the same penchant for flamboyant decorative motifs, later furnishings all but eschewed heavy ornamentation. Less expensive painted or stenciled decorations were preferred as the style developed. Later, the rich details that marked French Empire would disappear entirely in American Empire furniture, replaced by a new aesthetic that valued the quality of wood and carving over ornament.

Similarly, German craftsmen, artisans, and furniture makers took the Empire designs that were created for French nobles and adorned them with less bombastic ornamentation, creating Biedermeier, a style that became very popular in Germany, Austria, and parts of Scandinavia. This isn't to say that Biedermeier furniture was plain, even though *bieder* means just that in German. A typical Biedermeier table or chair was essentially an Empire piece that had been stripped of gilt ornamentation or other haughty detail (these had fallen out of fashion as Napoleon's power waned.) Yet perhaps to compensate for the lack of ornamentation used, Biedermeier craftsmen often created wildly incongruous—and impractical—shapes for many of the pieces.

With the emphasis on shape rather than on elaborate detailing, however, Biedermeier furniture is today considered one of the most versatile of Neoclassical styles. The Biedermeier predilection for elegant ebony inlays against exotic light woods make the pieces adaptable to interiors of any style, from Victorian to contemporary.

BELOW: *This hall doesn't have columns or pilasters to make it stand out as a Neoclassical interior. Rather, the checked floors, the Greek key design on the console, and the bronze statuary lend the space its unmistakably Neoclassical air.*

FURNITURE

ABOVE: *A modern interpretation of a classic Regency writing desk makes an attractive focal point for this vignette. Other classical elements combine seamlessly with Art Deco touches, such as the wallpaper design and the occasional chair.*

ABOVE: *The dark, rich woods of these Empire furnishings work in tandem with plain white walls to create a refined room setting. At the center, a round Biedermeier hall table features exaggerated cabriole legs and inlaid motifs. In the background, a sense of geometry is created with two Greek urns placed on sconces flanking a Federal-style mirror.*

OPPOSITE: *Rich with texture and pattern, this grouping calls on Neoclassical aesthetics to create a sense of timelessness. Maple, a favorite wood of craftsmen who fashioned furniture during the Empire period, loses none of its character when used in updated Biedermeier designs. Here, a contemporary rendition of Empire silhouettes uses maple to cast a golden glow on the dining room.*

FURNITURE

OPPOSITE: *Although the richly paneled wood walls are more characteristic of Victorian-era aesthetics, Neoclassical elements nonetheless sit well in this living room. Brightly colored draperies are swagged and drawn back in the typical Regency manner, while a gilded eagle—a favorite Napoleonic device—acts as a base for a console.*

ABOVE: *Bold colors and classical detailing create an opulent feeling in this room. A Georgian ebonized sofa—no doubt inspired by a design by American cabinetmaker Duncan Phyfe—provides the focal point, while accents such as the Greek bust atop an Ionic column and the gilt pediment console mirror support the Neoclassical ambience.*

FURNITURE

LEFT: *Placed away from the wall, this gilt-edged sofa, with its roll pillows, rectangular back, and nailhead trim, is distinctly Georgian in design. The sumptuous gold-colored jacquard silk of the upholstery contrasts well with other Neoclassical appointments in the room, including the corner niche curio cabinet, ebonized side chair, and gilt-framed oil painting.*

FURNITURE

ABOVE: *With its elegantly scrolled arms, unusual friezework decorating the frame, and off-white upholstery, this sofa bears the unmistakable mark of an Empire-inspired piece. The two trompe l'oeil oil paintings depicting blown-up decorative details draw attention to the lines of the sofa, which, with its casually arranged pillows, gives a shabby-chic accent to a Neoclassical form.*

FURNITURE

ABOVE: *This sunny yellow room is Neoclassical style at its most casual. Reflecting a typical classical trait, furnishings and accents are arranged in a proportional, geometric order. Fine details in the lyre-backed occasional chairs, as well as the silk upholstery fabric, pay homage to Empire style.*

RIGHT: *This interior is pure French Empire, with its fancifully painted furnishings. The trompe l'oeil pilasters, medallion, and friezework make up for any lack of classical architectural detail. The folding screen, silk fabric—swagged and draped over a plain vanity—and painted floor all point to Empire-inspired style.*

FURNITURE

OPPOSITE: *This living room's simple palette and attention to detail aptly demonstrate why Neoclassical remains a beloved style among modern designers. Disparate elements—two stand-alone fluted columns, plain slipcovers on the sofa and chair, a British Colonial–era occasional chair, and an ornate gilded mantel mirror—produce a surprisingly modern effect.*

ABOVE: *The mark of Neoclassical style is eminently recognizable here, with fluted columns acting both as support and as decoration. A granite tabletop is the perfect accompaniment, having the same cool, linear appearance as these classical elements.*

CHAPTER FOUR
ACCENTS

One of the defining characteristics of a French Empire–style room is the luxuriously abundant use of fabrics. Hung as multilayered curtains or draped in a tent-like fashion around a massive bed, brilliantly colored silks, tapestries, and velvets were often as much a part of the Neoclassical effect as furniture or architecture. In Napoleonic France, sumptuous textiles were a luxury that went hand-in-hand with the trappings of a Neoclassical interior. Plain painted walls were too severe for this style: hung with delicately patterned silk, walls came alive with the vibrant colors of the day, including bright yellows, greens, and reds.

Although the Baroque and Rococo periods lay claim to having the most extravagant details in furnishings and interiors, Neoclassical styles were hardly lacking. Fringes, tassels, and elaborate trim often adorned upholstered items and draperies in Empire settings. Of course, the designers of these elegant rooms would always balance elaborate details with the clean lines associated with classical furniture.

In England, designers had the same predilection for using rich fabrics, except that they more often used it as wallpaper, stretching fabrics such as silk damask, lustring (glazed taffeta), tabouret (half-silk in a plain color), or wool over a wall. The English did, however, use plenty of fabric for window treatments: heavy with fringing, draperies were often elaborately looped and swagged to accentuate the lines of the Regency furniture.

As on the walls, color and opulence are key factors in decorating floors in the Neoclassical style. Designers of the era often chose large carpets in rich colors with stylized repeat motifs, such as laurel wreaths, Greek keys, or hexagons. Aubusson rugs, which often feature hexagonal designs and floral motifs, are a favorite among fans of Neoclassical interiors. If designers chose to forego carpets or rugs, spectacular wooden or marble floors inlaid with contrasting details were usually the reason why.

Another detail that often received great attention in a Neoclassical setting was the mantel clock. Because fireplace mantels had become elaborate sculptures in and of themselves during the late eighteenth and early nineteenth centuries, designers came to view them as ideal

OPPOSITE: *Elegant and restrained, this grand salon depicts essential Empire style with its window treatments. As was the fashion in Napoleonic France, these tall windows are embellished with generous amounts of gold-colored silk; the fringed draperies are tied back with matching silk tassels and topped with extravagantly swagged pelmets.*

platforms to display collections or prized items, one of the most common being clocks.

Often gilded and displaying an allegorical scene or figure, these miniature sculptures were seen as works of art, created by clockmasters who took their craft very seriously. And no wonder—mantel clocks were generally placed at the center of the mantelpiece, altar-style, where they undoubtedly received many long glances. Icons of classical mythology—muses, gods, godesses, musical instruments such as lyres, and creatures such as chimeras and dragons—were the most popular choices for motifs. Otherwise, the designs were influenced by classical architecture, with clocks resembling miniature porticos, flanked by columns.

Clocks often featured gilding, one of the only stylistic elements carried over from the highly decorative Rococo period. This look, called ormolu, was also achieved with a copper-zinc alloy. Ormolu can also be seen on mantelpieces, mirrors, door frames, sconces, and candelabras produced during the Empire and Regency periods. For the most part, ormolu was eventually replaced by more delicate, understated accessories made of porcelain (although porcelain, too, was often detailed with gold).

Sèvres, the French national porcelain factory, told the story of Napoleon's conquests (and led the rage for French Empire–style porcelain) through its massive vases, urns, extensive dinner services, and plaques—all of which were based on ancient Greek, Roman, or

Egyptian pottery forms. Like furniture in the Empire style, Sèvres porcelain was richly decorated with scenic paintings and gilded details.

In England, Josiah Wedgwood had created an equally enduring legacy with his factory's production of cream-colored earthenware, which he dubbed Queen's ware, having secured royal patronage. This, along with the black Basaltes ware and the now-famous "Wedgwood Blue" Jasper ware, are as striking in their simplicity and form as Sèvres is complex and ornate.

Though many of the pieces are small, the porcelain created by Sèvres, Wedgwood, and others of the period is by no means insignificant: this decorative art form not only ties a Neoclassical room together; it is representative of the style itself.

A combination of two opposite aesthetics—one elaborately ornate, the other austere and reserved—Neoclassical style has inspired countless designers and architects, from the creators of French Empire style to those of Germany's Biedermeier style. To casual observers, it may be surprising that such different aesthetics are considered part of the same style family, but aficionados of Neoclassical understand that this duality is what makes the style unusually fresh—every time it's reborn.

OPPOSITE: *Applied to modern innovations such as lighting, Neoclassical style shines. Here, an amusing lamp has been fashioned using a marbleized Ionic column as its base and a plain geometric shade. Other accents intensify the classical look, including miniature Roman "ruins" as tabletop statuary and gold-colored silk drapery tied back with tasseled rope.*

ABOVE: *A bold stroke of Empire-inspired color turns a narrow hall into a show-stopper. The canary yellow of the walls is broken only by a trompe l'oeil molding effect, which is repeated in a fabric valance in the window. Carefully placed objects heighten the Neoclassical look, including a bust sitting atop a fluted column, a porcelain urn, and window-shade fabric displaying ancient Greeks in profile.*

ACCENTS

LEFT: *On this Empire-style balcony, simple square balusters support a thin handrail, as was the fashion in the late eighteenth and early nineteenth centuries. The scrolled balcony support bracket is distinctly Neoclassical in style, with its oval patera (a small oval or round ornament in classical architecture, often decorated with flowers or leaves).*

OPPOSITE: *Set against a pure white staircase, this bust and column pedestal capture the singular spirit of Neoclassical style. The geometric panels of the wall and richly carved balustrade and newel posts of the staircase point to a Georgian influence.*

ACCENTS

ABOVE: *This fireplace exemplifies the kind of architectural, Palladian-inspired refinement that marked Neoclassical designs of the English Regency. The overmantel, which directly imitates the design of the mantel, was an indication of wealth during the period, as was the display of china. Here, a few choice pieces add a touch of color to the otherwise austere white fireplace surround.*

ACCENTS

ABOVE: *Neoclassical can prevail in traditional settings even when it uses only modest elements of the style. In this living room, for example, the fireplace surround is a pared-down English Regency design, with a Wedgwood-inspired frieze—complete with Greek urn—decorating the lintel.*

ABOVE: *With flanking caryatids (busts on pillars) that support an elaborate frieze on the mantelpiece, this early Georgian fireplace is the centerpiece for a Neoclassical salon, which also features a Greek-temple-inspired doorframe and matching bookcase.*

RIGHT: *The fireplace often formed the visual centerpiece of a room during the Georgian period in England. Ornate fireplace surrounds, such as this example, reflect the great influence of classical style; the Greek key design, carved around the area meant to hold a mirror or painting, is the most obvious influence of antiquity. Elaborate friezework shows typical Neoclassical motifs, including ribbons, vines, florets, and swags.*

ACCENTS

ABOVE: *The striking contrast of cherry trim against the light paneled wood of this bath is an exmple of Biedermeier-inspired elegance. Biedermeier furniture was clean-lined, comfortable, and, at its best, charming. Here, panels of maple—a trademark wood of craftsmen of the period—create a warm, inviting look.*

ABOVE: *Built-in furniture came into vogue during the English Regency period, when architect Robert Adam greatly influenced interior design. Here, a modern take on classical architecture creates an entirely different tempo for this room. Modeled after a Greek temple, this built-in bookcase reflects the impeccable proportions and linear beauty for which Neoclassical style is famous.*

ABOVE: *The most commonplace features of a house are never neglected by devoted architects of Neoclassical style. Here, a post-modern interpretation of a Greek temple sits atop a door frame. Fluted glass panes emphasize the short pilaster details in the entablature, while the rich cherry wood lends a sense of warmth to the design.*

ACCENTS

LEFT: *This corner is comprised of singularly classical elements, including a Corinthian column with smooth marble shaft, several prints of Greek buildings, and a framed relief. The subdued color of the wall allows the furnishings to take center stage.*

OPPOSITE: *Arranged with an eye toward geometry and order, the urn prints hanging on the wall impart a strongly contemporary Neoclassical flavor to this living room. Underscoring this theme are the simple white-on-white upholstery, the decorative wall brackets, the Grecian urn–like vase and lamp, and the painted white cocktail table, which features dentil molding and pilaster detailing.*

ACCENTS

OPPOSITE: *Ancient Greece was no doubt the inspiration for this modern-day dining room. The scroll pattern of the Grecian urn—sketched along with the other examples of the art lining the walls—is repeated in the ceiling border. Bronze statuary and the room's overall neutral palette impart a refined, classical sophistication.*

ABOVE: *The "antiqued" effect of this hall sets the stage for a spare Neoclassical setting. The walls are painted to achieve the cool, detached look of stone, while a massive marble urn used as a planter brings the allure of an Italian garden to the space. An occasional chair lends a bit of French style.*

ACCENTS

ABOVE: *The illusion of trompe l'oeil, a technique used frequently by Neoclassical artisans, gives this vestibule architectural depth and character. Pictures of buildings sketched in the classical manner "hang" from tasseled ropes, which are suspended by brass rings; a "rope" defines both the chair rail and the molding.*

OPPOSITE: *Creating the illusion of Neoclassical grandeur, the painted effects on the walls of this living room point to Empire-style influence. Continuing the theme—in updated fashion—are the slipcovered camelback sofa and ottoman, the bright decorative pillows, and the slender Corinthian column casually leaning in a corner.*

ACCENTS

ABOVE: *Trompe l'oeil walls figure prominently in creating a Neoclassical setting for this modern bath. Other elements that contribute to the classical look include an ornate wall bracket that holds a Grecian urn, marble used generously throughout the room, a miniature marble statue and vase, and a candelabra, which gets its "ancient Roman" patina from oxidation.*

ACCENTS

ABOVE: *Heavily swagged and gathered on one side, the draperies in this hall hint at the Regency-style bathroom beyond its portal. The stripes of the occasional chair in the hall corner are echoed in the towels as well as in the upholstered seat of the X-bench, a classic Empire design.*

ABOVE: *Glazing, a technique in which bars hold the window panes in place, came into fashion during the early sixteenth century, long before Neoclassical style was all the rage. However, the simple geometry of diamond-pattern glazing works well with the classically swagged draperies of this bath. A niche—complete with urn—at the head of the tub heightens the timeless appeal of the room.*

ACCENTS

ABOVE: *Neoclassical details are reflected everywhere in this modern bath. Using a bold periwinkle background, the marble-frame mirror, which features a simple swag on its entablature, shows in its reflection other classical elements. For example, the overdoor has a modest broken pediment design, similar to door styles used throughout the Neoclassical period in England.*

ABOVE: *Though not architecturally distinctive, this bathroom combines classical motifs with a neutral palette to create a Neoclassical mood. The nailhead trim at the base of the slipper chair is fashioned in a Greek key design—a quintessential Neoclassical motif—while the high wainscoting offsets a lightly contrasting geometric pattern in the wallpaper.*

OPPOSITE: *Even without the other elements that indicate the style, this bath's black-and-white palette is clearly classical. From the granite sink top to the marble tiles to the heavily draped shower curtain—which is tied back in Neoclassical fashion—the room's colors speak volumes. A wood Empire-style vanity mirror lends an additional stylistic touch.*

INDEX

Accents, 75–95. *See also specific styles*
  Empire-style, 75
  English, 75
  fabrics, 75
  mantel clocks, 75–76
  porcelain, 76–77
Adam, Robert, 13, 20–21, 85
American Empire style, *See* Federal style
Architecture, 19–39. *See also specific styles*
  classical, 19–20
  Georgian, *18,* 19, 21
  Greek Revival, 20–21, *22*
  Neoclassical, 20–21
  Palladian, 20–21, *21*
  Regency, 20

Balconies, 78, *78*
Bathrooms, 38, *38–39,* 55, *55, 56,* 56, *58–59,* 59, 84, *84, 92–94, 92–95*
Bedrooms, 54, *54, 57,* 57
Biedermeier style, 13, 37, 62–63, 84
  examples, *64–65,* 84
Bookcases, 85, *85*
Busts, 82, *82*

Cabriole legs, *59,* 60
Caryatids, 82, *82*
Ceiling plasterwork, 21, *21*
Classical architecture
  basic orders, 19
  revival styles, 20–21
Clocks, mantel, 75–76
Collection of Designs for Household Furniture and Interior Decoration (Smith), 42
Color, *32,* 33–34, 52, *53,* 76, *76,* 86, *86*
Columns, 13–14, 19–20, 22, *22,* 26–27, 31, 34, *56, 72–73*
  Corinthian, *24–25, 25, 26,* 86
  Doric, *8,* 19, *30,* 50
  Ionic, *14,* 19, *48, 57*
  Tuscan, *15–16, 18,* 19, *22, 26–27, 29, 31, 33–34, 38, 47*

Conservatories, *8, 9*
Contrast among elements, 29, *29, 32,* 33
Corners, *48,* 86

Dining areas, 10, *10–11, 17, 34–36,* 37, *49, 51, 53, 65, 88, 89*
Doors, French, 29, *29*
Doorways
  arched, 30, *30–31,* 31, 34, *34*
  Empire, 42, *43*
  modern Neoclassical, 85, *85*
Draperies, *66,* 67, *74,* 75, *76, 93,* 93

Empire style, 9–10, 16, 41–42, 44, 59, 61–62, 70, 75–78
  American, *See* Federal style
  examples, *17, 30,* 40, *43, 45,* 50, *53–54, 59, 62, 69–71, 78, 91, 94*
English Regency, *See* Regency style
Entryways, *See* Foyers

Fabrics, 75
Federal style, 13–14, 62
  example, *64*
Fireplaces, 75–76, 80–83, *80–83*
Floors, 75
  geometric paving, *32,* 32, *37,* 37
  late Georgian, 44, *45*
  painted, *71*
Fontaine, Pierre-François-Léonard, 10, 41, 61
Foyers, 28–30
French doors, 29, *29*
French Empire, *See* Empire style
Furniture, 61–73. *See also specific styles*
  arrangement, 42, *47,* 47, *70,* 70
  Biedermeier, 62–63
  built-in, 85, *85*
  Empire, 61
  Federal, 62
  Regency, 61–62

Gardens, *24–25,* 25, *27,* 27
Gazebos, *24,* 25
Georgian style, *18,* 19, 33–34, *35,* 44, *45,* 67, 68, *68,* 82, *82–83*
Gilding, 76
Greek Revival style, 20–21, *22*

Halls, *63,* 77, 89, *89*
Hepplewhite, George, 14
Holland, Henry, 42
Hope, Thomas, 42
Household Furniture and Interior Decoration (Hope), 42
Houses, *18,* 19–20, 22, *22–23*

Interior design, 41–59. *See also specific styles*
  Empire style, 41–42
  Regency style, 42

Key motif, 20, *83,* 94
Kitchens, 37, *37*

Lamps, *76,* 77
Latrobe, Benjamin Henry, 20
Living rooms, *47, 50,* 67, *72, 81, 87, 91*

Mantels, *See* Fireplaces; Clocks; Overmantels
Marble, 28, *28, 92,* 92
Modern design, with Neoclassical tones, 28–29, *36, 46, 48–49, 72, 76*

Napoleon Bonaparte, 9–10, 41, 76
Neoclassical style, 9–14
  architecture, 20–21

Obelisks, 28, *28*
Ormolu, 76
Overmantels, 80, *80*

Palladian design, 20–21, *21, 80*
Patios, *26,* 26
Percier, Charles, 10, 41, 61
Pergolas, *27,* 27
Pilasters, 19–20, 38, *39*
Porcelain, 76–77

Regency style, 10–13, 42, *47,* 61–62, 75, 80–81, 85
  examples, *16, 47, 52, 64, 66, 80–81, 85, 93*
Sèvres, 76–77
Sheraton, Thomas, 14
Smith, George, 42
Sofas, 52, *52, 60,* 61, 67–69, *67–69, 91*
Sphinxes, 16, *16, 62, 62*
Staircases, 78, *79*

Trompe l'oeil, 44, *44, 58–59,* 59, *69, 71, 90, 90, 92, 92*

Verandas, *26,* 26

Walls, *32,* 33–34, *35, 38, 39,* 75, *79, 86–88,* 89–90, *90–92,* 92, *94. See also* Trompe l'oeil
Wedgwood, 77
Windows
  glazing, *93,* 93
  Palladian, *21*
Window treatments
  Empire-style, *74,* 75
  Regency, 75
Woods, 62, *64,* 64, 85, *85*

X-benches, *12,* 59, *59,* 93

PHOTO CREDITS

©Antoine Bootz: 12 (designer: Mimi Russel), 28 (designer: Ira Levy), 49 (designer: Vicente Wolf), 64 left (designer: Karl Kemp), 86 (Mimi Russel)

©Steve Gross and Susan Daley: 2 (designers: Tony Inson and Alan Hicklin), 8 (designer: Michael Trapp), 15, 30 left, 52, 64 right, 66 (designer: Michael Valente), 67 (designers: Tony Inson and Alan Hicklin), 74, 84

©Nancy Hill: 29 left (decorating and remodeling courtesy of *House Beautiful,* Stylist: Judith Driscoll)

©image/dennis krukowski: 24 (Rye Interior Visions Showhouse), 70 (designer: Lincoln Interiors), 71 (designer: Michael Tyson Murphy), 76 (designer: Michael Christiano), 90 (decorative painting: Michael Lane)

©Keith Scott Morton: 16, 17, 51, 54, 79, 80, 82, 93 both

©Robert Perron: 22, 44, 85 both

©David Phelps: 94 right (courtesy *American Homestyle* magazine, designer: Michael Berman)

©Paul Rocheleau: 10–11, 20, 21 both, 29 right (architect: Michael Graves), 35, 36 (architect: Michael Graves), 43, 53, 78, 83

©Eric Roth: 5 (architect: Robert Miklos of Schwartz Silver Architects), 14 left, 27, 31 (designer: Greg Cann of Cann & Co.), 37 (architects: Olson Lewis Architects), 39 (designer: Peter Labau of Classic Restorations), 46, 47 (designer: Carole Kaplan of Two by Two Interior Design), 48 left (designer: Ron Bower Studios), 48 right, 50 (designer: Brad Morash Interior Design), 63, 65 (designer: Douglas Truesdale of Geib Truesdale Interior Design), 72 (designer: Peter Wheeler of PJ Wheeler Associates), 81 (designer: Wendy Reynolds of Cheever House Decorations), 87 (designer: Ken Kelleher), 91 (designer: Jim Anderson of Anderson Glass Arts)

Studio South II/©Ken Murphy: 18, 23, 26 left

©Tim Street-Porter: 40 (designer: George Gotti), 88 (designer: Thomas Beeton)

©Jessie Walker Associates: 30 right

©Elizabeth Whiting Associates: 7, 13, 14 right, 25, 26 right, 32, 33, 34, 38, 45, 55, 56, 57, 58, 59 both, 60, 62, 68, 69, 73, 77, 89, 92, 94 left, 95